KEY CONCEPT ACTIVITIES

PREALGEBRA
EIGHTH EDITION

Elayn Martin-Gay
University of New Orleans

The author and publisher of this book have used their best efforts in preparing this book. These efforts include the development, research, and testing of the theories and programs to determine their effectiveness. The author and publisher make no warranty of any kind, expressed or implied, with regard to these programs or the documentation contained in this book. The author and publisher shall not be liable in any event for incidental or consequential damages in connection with, or arising out of, the furnishing, performance, or use of these programs.

Reproduced by Pearson from electronic files supplied by the author.

Copyright © 2019 by Pearson Education, Inc.
Publishing as Pearson, 501 Boylston Street, Boston, MA 02116.

All rights reserved. No part of this publication may be reproduced, stored in a retrieval system, or transmitted, in any form or by any means, electronic, mechanical, photocopying, recording, or otherwise, without the prior written permission of the publisher. Printed in the United States of America.

1 18

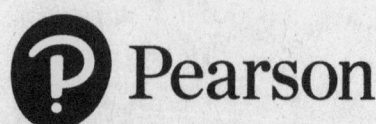

ISBN-13: 978-0-13-470872-0
ISBN-10: 0-13-470872-5

Key Concept Activity Lab Workbook, *Prealgebra* 8e

Table of Contents

Chapter 1—The Whole Numbers
1. Extension Exercise—Hidden Costs of Planning a Party ... 1
2. Exploration Activity—Comparing Economic Data ... 3
3. Conceptual Exercise—Safe Dose ... 5
4. Group Activity—Where Did All the Money Go? ... 7

Chapter 2—Integers and Introduction to Solving Equations
1. Extension Exercise—Temperature Scales ... 9
2. Exploration Activity—Balancing Your Account ... 11
3. Conceptual Exercise—Rent-a-Calculator ... 13
4. Group Activity—Elevator Talk ... 15

Chapter 3—Solving Equations and Problem Solving
1. Extension Exercise—Mystery Dose ... 17
2. Exploration Activity—Community Fund-Raiser ... 19
3. Conceptual Exercise—Money in Trees ... 21
4. Group Activity—Springs in the Ozarks ... 23

Chapter 4—Fractions and Mixed Numbers
1. Extension Exercise—Monitoring Your Stocks ... 25
2. Exploration Activity—Bills and Coins ... 27
3. Conceptual Exercise—Team Effort ... 29
4. Group Activity—Cubes and Fractions ... 31

Chapter 5—Decimals
1. Extension Exercise—A Year at MIU ... 33
2. Exploration Activity—Exchange Rates ... 35
3. Conceptual Exercise—Traveling for Big Brother ... 37
4. Group Activity—The Donut Dilemma ... 39

Chapter 6—Ratio, Proportion, and Triangle Applications
1. Extension Exercise—Cost of Living ... 41
2. Exploration Activity—A Trip to the Ozarks ... 43
3. Conceptual Exercise—The UFO Files ... 45
4. Group Activity—Eating Well ... 47

Chapter 7—Percent
1. Extension Exercise—Batting Champ ... 49
2. Exploration Activity— A Diverse India ... 51
3. Conceptual Exercise—Where's the Beef? ... 53
4. Group Activity—M&M's® ... 55

Chapter 8—Graphing and Introduction to Statistics and Probability
1.	Extension Exercise—Winning BigBucks Lottery	57
2.	Exploration Activity—What Is Average?	59
3.	Conceptual Exercise—Lightning Strikes	61
4.	Group Activity—Laundry Blues	63

Chapter 9—Geometry and Measurement
1.	Extension Exercise—Survival of the Hatchlings	65
2.	Exploration Activity—Fence Me In	67
3.	Conceptual Exercise—Converting Doses	69
4.	Group Activity—Composting	71

Chapter 10—Exponents and Polynomials
1.	Extension Exercise—Shipping Box of Maximum Volume	73
2.	Exploration Activity—Computing Distances	75
3.	Conceptual Exercise—Deaths from Motor Vehicle Crashes	77
4.	Group Activity—Worldwide Internet Users	79

Answer Key
Answers are available. Talk to your instructor.

1. **Extension Exercise—Hidden Costs of Planning a Party**

Your company is planning an employee holiday party. In addition to the direct costs, such as food and door prizes covered by the budget, the company president wants to know hidden labor costs, or how much the party will cost the company in terms of planning time. You need to help the party committee calculate the hidden labor costs for planning the party. The four employees on the party committee are Karen, Tom, Juan, and Debbie.

a. After subtracting vacations and holidays for the year, Karen, Tom, Juan, and Debbie, each work 48 weeks a year, 40 hours a week. Determine how many hours each employee works in a year.

Hours worked per year per employee: _____

b. The annual salaries for Karen, Tom, Juan, and Debbie are shown in the table below. For each person, determine the hourly cost (in dollars) to the company, or how much one hour of work costs the company. Show your answers in the table.

Hint: hourly cost = yearly pay ÷ number of hours worked.

Employee	Yearly Pay ($)	Calculations	Hourly Cost to Company ($)
Karen	$28,800		
Tom	$30,720		
Juan	$32,640		
Debbie	$36,480		

c. Karen, Tom, Juan, and Debbie met five different times for two hours each to plan the party. Each meeting was held around lunchtime, so that one hour of the meeting was during the lunch hour (not on company time) and the other hour was on company time. How many hours of company time did each committee member spend planning the party? Show your work.

Hours of company time each committee member spent planning: _____

d. In the chart below, copy the hourly cost for each employee (from part *b*). For each of the committee members, record the number of planning hours on company time (from part *c*), and determine how much the planning meetings cost the company in lost wages.

Employee	Hourly Cost ($) (copy from part *b*)	Hours on Company Time (copy from part *c*)	Calculations	Cost to Company ($)
Karen				
Tom				
Juan				
Debbie				

e. Explain why the direct costs of a party, such as food and drinks, decorations, and prizes do not necessarily cover all of the costs when a company plans a party for its employees. Explain what other hidden or direct costs the company might have if the party is held on company property or during regular work hours.

_____

_____

_____

_____

_____

_____

2. Exploration Activity—Comparing Economic Data

The information in the chart below is taken from the U. S. Census Bureau. It compares economic information for three industries in the United States for the years 2007 and 2012.

Type of industry	Annual payroll (in dollars)		Paid Employees	
	2007	2012	2007	2012
Mining	40,687,472,000	59,461,950,000	730,433	848,189
Hospitality	170,826,847,000	196,103,341,000	11,600,751	12,007,689
Health Care and Social Assistance	662,719,938,000	801,239,522,000	16,792,074	18,414,757
Total				

a. For 2007 and 2012, find the total annual payroll (in dollars) and the number of paid employees in the mining, hospitality, and health care/social assistance industries. Record results in the table above. Use the vertical format of the numbers to help you calculate the totals for each column.

b. Now determine the increase in annual payroll from 2007 to 2012 and the increase in the number of paid employees from 2007 to 2012. Complete this part by hand or by using a scientific calculator. Write your results in the table below. *Hint:* The increases are actually differences.

Type of Industry	Increase in Annual Payroll (in dollars)	Increase in Number of Paid Employees
Mining		
Hospitality		
Health Care and Social Assistance		

Chapter 1: The Whole Numbers

c. For each type of industry in your table from part b, round the increase in annual payroll to the nearest million dollars. Then round your data for the increase in number of paid employees to the nearest hundred thousand. Add two columns to your table from part b and enter your results from part c in these two new columns.

d. Describe the process that you would use to find the average salary for each industry in 2012. Which data would you use? Determine whether or not your results would be whole numbers.

_____

_____

_____

_____

3.   Conceptual Exercise—Safe Dose

Claudia, a 7-year old girl, weighs 19 kilograms. The manufacturer's recommended safe range for doses of heparin is 10 to 25 units per kilogram per hour. Through a series of steps, you will determine the safe range of doses for Claudia and determine her dose of heparin per hour.

a. 10 units of heparin per kilogram per hour is the low end of the recommended safe dose. What is the lowest safe dose of heparin that should be given to Claudia in one hour?
Hint: The lowest safe dose (number of units) of heparin for Claudia in one hour = the lowest safe number of units of heparin recommended per kilogram per hour × Claudia's weight in kilograms.

Lowest safe dose of heparin for Claudia in one hour: _____ units

b. 25 units of heparin per kilogram per hour is the high end of the recommended safe dose. Calculate the highest safe dose of heparin per hour for Claudia.

Highest safe dose of heparin for Claudia in one hour: _____ units

c. To summarize the results in parts *a* and *b,* record the recommended safe range (low to high) of units of heparin per hour for Claudia.

From _____ to _____ units of heparin per hour

d. Claudia must receive her dose of heparin as part of an IV or intravenous fluid. The total amount of the IV solution (containing her dose of heparin) is 250 milliliters, and it flows at a rate of 50 milliliters per hour. How many hours will it take to administer 250 milliliters of the intravenous (IV) solution containing heparin? Show your work.

_____ hours

e. Suppose that the total IV solution (250 milliliters) contains 2000 units of heparin. How many units of heparin are administered to Claudia per hour? Show your work.

Hint: The number of units of heparin administered to Claudia per hour = total number of units of heparin in the entire 250-milliliter solution ÷ the number of hours it would take to administer the entire solution (from part *b*).

_____ Units of heparin administered to Claudia per hour

f. Compare your answer in part *e* to the safe range of dosages for Claudia as determined in part *c*. Is this dose within the recommended safe range? Explain why or why not.

_____

_____

_____

_____

_____

4. Group Activity—Where Did All The Money Go?

Mary Thurmond willed her entire estate to be divided equally among her 22 nieces and nephews. When she died, her estate was valued at over two million dollars. First, in a very simplistic world where there are no fees taken out of the estate, you will determine the amount of inheritance for each of the nieces and nephews. Next, you will determine how to distribute the funds after taxes and probate (the act of verifying the will's validity) fees are subtracted from the estate.

Work with your group members to answer the following questions.

a. In addition to her residence, Mary Thurmond owned a beach house and a farm that was leased to a corporate farmer. Her real estate properties were appraised as shown in the chart below. Calculate and record the total appraised value of her real estate properties in the table.

Property	Appraised Value of Real Estate
Residence	$435,000
Beach House	$350,000
Farm	$730,000
Total	

b. An estate auctioneer estimated the value of her personal belongings at $115,000. Investments in money market accounts and stock market accounts are estimated at $573,000. Find the total estimated value of her personal (not real estate) belongings and investments.

c. What is the total estimated value of all of Mary Thurmond's properties, belongings, and investments? Round this amount to the nearest hundred thousand dollars.

d. Work with your group members to calculate each nephew's and niece's inheritance based on your final rounded estimate in part *c*.

e. An auction was held to sell Mary Thurmond's personal belongings. The real estate properties and stocks were sold, and the money market accounts were closed out. All of this money was held in an escrow account. (Note: An escrow account is money or property placed into the hands of a third party, and is delivered to the grantee only after fulfilling specific conditions.)

The money collected after all real estate, broker, and auctioneer fees were paid is shown in the table below. Find the difference between the actual dollar amount collected after fees and the appraised value. Then determine whether the actual dollar amount after fees reflects a gain or loss in value from the appraised estimate.

Properties	Actual Amount Collected After Fees	Difference between Actual Amount and Appraised Value	Gain or Loss in Value?
Residence	$394,190		
Beach house	$360,100		
Farm	$522,000		
Personal belongings	$53,250		
Stocks and Money Market	$427,930		
Total			

What is the total amount of money that was deposited into the escrow account?

f. Calculate the share of the "total in escrow" for each nephew and niece. Find the difference between this share and the estimated share based on appraised values (from part d).

g. Probate court fees came to $55,000, and taxes totaled $575,230. What total sum was left to divide among the heirs after these costs are paid from the escrow account?

h. After taxes and probate fees are paid, how much money will finally be given to each nephew and niece? Round to the nearest cent.

i. Calculate the difference between what each niece and nephew might have expected given the appraised values (from part d) and their actual share (from part h).

Key Concept Activity Lab Workbook, *Prealgebra* 8e Chapter 2: Integers and Introduction to Solving Equations

1. **Extension Exercise—Temperature Scales**

   If C is the temperature in degrees Celsius, and F is the temperature in degrees Fahrenheit, then we can represent the relationship between temperature scales with the following formula.

   $$C = \frac{5}{9} \cdot (F - 32)$$

   a. Convert the Fahrenheit temperatures of 32°F, 23°F, and 14°F to the Celsius scale. Use the two-step process described next to arrive at your answer.

   Step 1: Subtract 32 from the given temperature.

   Step 2: Multiply by $\frac{5}{9}$.

   $C = \frac{5}{9} \cdot (32 - 32)$ $C = \frac{5}{9} \cdot (23 - 32)$ $C = \frac{5}{9} \cdot (14 - 32)$
   $C =$ $C =$ $C =$

   b. Let's reverse the above procedure by first describing the opposite of Step 2 and then the opposite of Step 1. This new two-step procedure will allow you to convert any Celsius temperature to Fahrenheit temperature.

   Reverse of Step 2: _____

   _____

   Reverse of Step 1: _____

   _____

   c. Show how to convert the Celsius temperatures of 0°C, −5°C, and −10°C to Fahrenheit temperatures using the reverse procedure developed in part *b*.

d. Letting C be the temperature in degrees Celsius and F be the temperature in degrees Fahrenheit, suggest a formula for converting from degrees Celsius to degrees Fahrenheit.

e. Temperatures in space are measured in degrees Kelvin. Find the relationship between the Kelvin scale and the Celsius scale by observing the last two columns.

Temperature of	Kelvin	Celsius
Earth's Sun	6000	5727
Water Boiling	373	100
Water Freezing	273	0
Absolute Zero	0	−273

- What operation can you perform to convert a Kelvin temperature to a Celsius temperature?

- What operation can you perform to convert a Celsius temperature to a Kelvin temperature?

f. If K is a temperature in degrees Kelvin and C is a temperature in degrees Celsius, then suggest a formula for converting degrees Kelvin to degrees Celsius and a formula for converting degrees Celsius to degrees Kelvin.

2. Exploration Activity —Balancing Your Account

You just received your bank statement, which shows an end-of-month balance of $25 in your checking account. Where did all the money go? You decide to compare last month's deposits and withdrawals. A good way to do this is to take the balance from the start of last month. Then add a positive number for deposits, and add a negative number for withdrawals. The following table shows this process, and the first calculation is completed for you.

Transaction	Withdrawal (−) ($)	Deposit (+)	Calculation	Balance ($)
Start of Month				+625
Rent	-550		+625 + (-550)	+75
Car Loan	-250			
Hair Cut	-25			
Paycheck		+600		
Auto Repair	-575			
Electric Bill	-75			
Telephone Bill	-110			
Cable	-25			
Paycheck		+600		
Groceries	-190			
End of Month				+25

(↑ for part *d*)   (↑ for part *e*)

a. Finish filling in the table by showing the calculation needed to obtain the balance after each transaction. Place the result of each calculation in the balance column as either a positive or negative number. Complete all calculations by hand, and then go back to check your work with a calculator.

b. Observe all the calculations where you added a positive number and a negative number.

- If you temporarily ignore the signs of the numbers, then what operation did you actually perform on the two numbers in order to obtain the correct balance?

- After performing the operation, what about the two numbers determined whether the balance was a positive number or a negative number?

- In your own words write a general procedure for adding positive and negative numbers.

_____

_____

_____

_____

c. Now observe the calculations where you added two numbers with the same sign (either both positive or both negative).

- If you temporarily ignore the signs of the numbers, then what operation did you actually perform on the two numbers?

- After performing the operation, what about the two numbers determined whether the balance was positive or negative?

- In your own words write a general procedure for adding two numbers with the same sign.

_____

_____

_____

d. Return to the table and add up all the withdrawals in the second column. Record the sum below and in the table (*End of Month* row, *Withdrawal* column).

e. Return to the table and add up all the deposits in the third column. Record the sum below and in the table (*End of Month* row, *Deposit* column).

f. Analyze the results of parts *d* and *e*. Then explain what is happening to your cash flow over this one-month period.

_____

_____

g. How did you manage to end up with +25 dollars when more money was going out than coming in?

_____

_____

h. Assuming you have no auto repairs or hair cuts next month, estimate your ending balance for next month. Show all your work.

3. **Conceptual Exercise —Rent-a-Calculator**

Suppose you decide to start a business renting graphing calculators to other students at your college. Before you can begin your business, you need to purchase some graphing calculators. You have enough money saved to buy 50 reconditioned calculators from a wholesaler for $45 per calculator. On campus, you advertise that any student can rent a calculator for one semester at a fee of $15 per semester, but it must be returned in good condition. You plan on running this business over a two-year period that will cover six semesters (fall, spring, and summer of each year).

a. What is the initial *cost* to start your business? Show your work.

b. If you manage to rent all 50 calculators for two consecutive semesters, what will be your *revenue* at the end of this time period? Revenue is the money coming in from the rentals. Show your work.

c. Use your results from parts *a* and *b* to calculate the *profit* after two semesters. Show your work.

$$Profit = Revenue - Cost$$

d. If the *profit* is negative, then explain the status of the business in terms of revenue and cost. Use complete sentences in your explanation.

_____

_____

_____

_____

e. How many calculators must you rent during your third semester of business to break even? Show all work.

f. By the end of the third semester, your business does break even. Now you set the goal of making $3000 profit over the next three semesters. To accomplish your goal, what is the new minimum rental price that must be charged for the calculators per semester?

g. What type of risk are you taking by charging a price greater than $15? Explain using complete sentences.

_____

_____

_____

h. Suppose the increase in price causes a decrease in demand such that you are only able to rent 40 calculators in each of the last three semesters of running your business. How much are you short of your $3000 goal if you use the rental price from part *f*?

4. Group Activity—Elevator Talk

Suppose you work in a building that has three floors above ground level and two floors below ground level. Imagine that the figure below represents an elevator that takes people up and down to different floors. Assume that the distance between successive floors is 10 feet.
Work with your group members to answer the following questions.

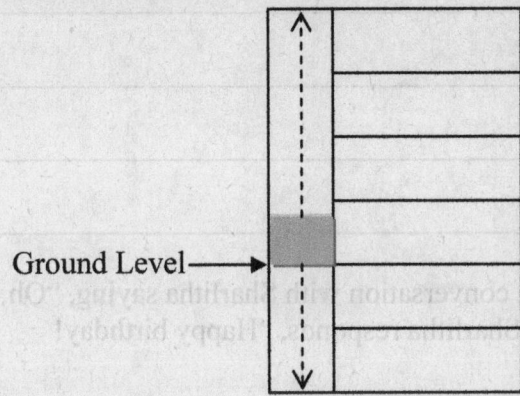

Ground Level

a. Label the ground level floor as 0, and then label the floors above floor 0 with the positive integers 1, 2, 3, and 4 and below floor 0 with the negative integers −1 and −2.

b. Sharlitha enters the elevator at ground level, travels up to floor 1 to drop off a letter, and then goes down to floor −2 to pick up some supplies. To the right of the figure above, draw and label two arrows pointing up or down to visualize how positive and negative numbers can represent her vertical movement on the elevator.

c. Use addition of positive and negative integers to show the calculation that represents Sharlitha's elevator trip and the solution that represents the floor she ends up on (floor −2).

d. Let 10 represent the number of feet Sharlitha traveled up and −30 the number of feet she traveled down. Use subtraction to represent the total distance that Sharlitha traveled between floors.

e. After picking up supplies on floor −2, Sharlitha gets back on the elevator with two other people, Terry and Marco. Marco says to Terry, "I do not understand how these floors are numbered. If zero means *nothing*, then it's impossible for anything to be less than nothing." State whether you agree or disagree with Marco and support your answer in complete sentences.

___

___

___

___

f. Terry gets out on floor −1. Then Marco starts up a conversation with Sharlitha saying, "Oh, I have the money I borrowed from you yesterday." Sharlitha responds, "Happy birthday! Consider that money as my treat for lunch."

If b represents the amount of money that Marco borrowed from Sharlitha, then how would you interpret the following rule? $-(-b) = b$ $(b > 0)$

Hint: The great mathematician Leonard Euler once said, "...*to cancel a debt signifies the same as giving a gift."*

___

___

___

g. Marco and Sharlitha leave the elevator at floor 0. Before going back to work, Marco describes his five-month battle trying to lose weight. "In the first month of dieting, I cheated all the time and gained 10 pounds. During the second month, I lost 5 pounds. Then, in the third month, I lost 7 pounds. During the fourth month, I lost 3 pounds, and now I have gained 4 pounds during the fifth month." Marco refers to his weight journal below:

Month	January	February	March	April	May
Weight Change	+10	−5	−7	−3	+4

- Show the calculation needed to find the total change in weight over the 5-month period.

- How much did Marco gain or lose? Express your answer as a positive or negative integer.

1. Extension Exercise—Mystery Dose

It is difficult to calculate the exact surface area of a human body since it is not flat or shaped like a box. However, health professionals have come up with a formula to approximate the number of square meters it would take to cover a human body. All you need to determine a person's body surface area (BSA) is that person's height and weight. Once calculated, a person's BSA can be used to determine the proper individual dose for some medications.

The chart below contains data on two people. In each case, either the BSA or the proper dose is unknown and has been replaced by a variable, x or y.

Person	Body Surface Area (square meters)	Recommended Rule for Finding the Proper Dose	Proper Dose (milligrams)
Linda	1.5	3.7 mg. per BSA of 1 sq. meter	x
Jim	y	9.3 mg. per BSA of 1 sq. meter	23.3

a. For Linda, label what x represents in words by completing the "Let $x =$" statement below. Then write a general formula for Linda by using the known values to find the value for x. Round results to the nearest tenth of a milligram. Show your work below.

Let $x =$ _____

General formula for Linda: $x =$ _____

b. For Jim, label what y represents in words by completing the "Let $y =$" statement below. Then write a general formula for Jim by using the known values. If needed, solve for y. Round results to the nearest tenth. Show your work below.

Let $y =$ _____

General formula for Jim: $y =$ _____

c. For the next five days, Linda's dose needs to slowly increase. Use Linda's BSA from the chart on the previous page and the table below to find the proper dose (in milligrams) for Linda on each day. Round results to the nearest tenth of a milligram. Show calculations and record answers in the table below.

Milligrams per 1 square meter of BSA	Calculations for Linda	Amount of Dose (milligrams)
First day: 3.7		
Second day: 5.5		
Third day: 7.4		
Fourth day: 9.3		
Fifth day: 11.1		

d. Jim's dose each day needs to slowly decrease for the next five days. Use Jim's BSA from your answer in part *b* and the table below to find the proper dose (in milligrams) for Jim on each day. Round results to the nearest tenth of a milligram. Show calculations and record answers in the table below.

Milligrams per 1 square meter of BSA	Calculations for Jim	Amount of Dose (milligrams)
First day: 9.3		
Second day: 8.6		
Third day: 7.5		
Fourth day: 5.4		
Fifth day: 3.5		

2. Exploration Activity—Community Fund-Raiser

Your community is hosting a fund-raiser for charity. Decorations, pamphlets, and other miscellaneous items have been donated. Volunteers have planned the event and will prepare the food, so there are no labor costs for the event. It will cost about $5.25 per person to buy the food, and $320 to rent a hall for the evening.

a. Do the necessary calculations and fill in the following table to show how much the event will cost depending upon the number of people attending in the first column.

No. of People Attending	Cost of Hall Rental ($)	Cost of Food, $5.25 per person ($)	Total Cost ($)
50			
100			
150			
200			
250			
300			
350			

b. Write an expression that represents the cost of food if n people attend the fund-raiser.

cost of food for n people: _____

c. Write an expression that represents the total cost if n people attend the fund-raiser.

total cost for n people: _____

d. If C represents total cost and n represents the number of people attending the fund-raiser, write an equation that relates C and n. *Hint:* Use your expression from part c.

equation: _____

e. If people are charged $15 each for the dinner, calculate the total revenue for the dinner. Then copy the total cost from the table in part a into the 3rd column, and calculate the profit, which is revenue minus total cost.

No. of People Attending	Total Revenue ($) (at $15 per person)	Total Cost ($) (from part a)	Profit ($) = Revenue − Cost
50			
100			
150			
200			
250			
300			
350			

f. If R represents revenue and n represents the number of people attending the fund-raiser, then write an equation for R in terms of n.

$R =$ _____

g. Record your equations from parts d and f in the spaces provided below.

$R =$ _____ $C =$ _____

h. The profit of the fund-raiser will be the revenue minus the cost. Use the formulas in part g to write an expression for profit P. Then simplify the expression by removing parentheses and combining like terms.

$P =$ _____ − (_____) = _____
     Revenue   Cost   Simplified

i. Use the equation for profit in part h to determine how many people would have to attend the event to make a profit P of $800. Round your answer to the nearest whole person.

3. **Conceptual Exercise—Money in Trees**

When determining the dollar value of a tree, you must estimate the size of the tree. To do this, you need to calculate the tree's **diameter breast height (dbh)**, which is the diameter of the tree trunk at 4 ½ feet above the ground (breast-high on a man of average height).

a. There is no real monetary value of a tree until it obtains a diameter breast height (dbh) of about 14 inches. Use the formula, $C = \pi d$, to find the minimum circumference C (at breast height) for a tree to begin to have real monetary value. Round your answer to the nearest whole inch. *Hint:* $d = 14, \pi \approx 3.14$

Minimum circumference: $C =$ _____ inches

b. For each circumference given below, find the diameter breast height of the tree. First, substitute the given value for C into the equation, $C = \pi d$. Then solve the equation for d. Show your substitution and work in the second column, and record your answer for d in the last column. Round d to the nearest inch. *Hint:* $\pi \approx 3.14$

Circumference at Breast Height (inches)	Equation, Show work	What is *d*? (rounded to nearest inch)
47		
53		
66		
88		

c. If you know the circumference of a tree, how would you solve the equation $C = \pi d$ for d (so that it begins as $d = ?$) To check your equation, use it to calculate the diameter breast height d for each circumference in part b and see if you get the same answer.

$d =$ _____

d. If a tree's diameter decreases 1.5 inches every 12 feet up the tree (on average), then how much will the tree decrease in diameter for each 1 foot increase in height? *Hint:* You can set up and solve a proportion to figure this out.

Decrease in diameter per foot = _____ inches

e. Using the answer from part d, explain how you would find the decrease in diameter of a tree at 20 feet.

_____

_____

_____

f. Generalize the process that you used in part e and write a formula for the decrease in the diameter over x feet.

Decrease in diameter = _____

4. **Group Activity—Springs in the Ozarks**

Mark, a geologist, recently traveled to the Ozarks in Missouri and Arkansas. He is fascinated by the constant flow of underground springs that feed the many clear, deep pools in the area, and thus, he gathered the following information about two of these underground springs: Round Springs and Big Springs, both in Southern Missouri.

Name of Spring	Average Amount of Water Flow Per Day (millions of gallons)	Maximum Water Flow Ever Measured Per Day (millions of gallons)
Big Spring	276	800
Round Spring	26	300

The formula used to calculate the flow rate of a spring is $A = r \cdot t$, where A is the amount of water flow, r is the rate of water flow, and t is the measured amount of time. In the table, t equals one day. Remember that when using formulas, the units must match.

Work with your group members to answer the following questions.

a. Solve the formula above, $A = r \cdot t$, for r.

b. Rates are measured in such units as miles per hour, feet per second, gallons per minute, and kilograms per hour. At the springs, what unit is used to measure the flow rate of the water?

c. What is the average flow rate r for Big Springs? Use the data in the 2nd column of the table and the formula for r from part *a*.

Average flow rate, Big Springs: $r = $ _____ millions of gallons of water per day

d. What is the average flow rate r for Round Springs? Use the data in the 2nd column of the table and the formula for r from part *a*.

Average flow rate, Round Springs: $r = $ _____ millions of gallons of water per day

e. What is the maximum flow rate r for Big Springs? Use the data in the 3rd column of the table and the formula for r from part a.

Maximum flow rate, Big Springs: $r =$ _____ millions of gallons of water per day

f. What is the maximum flow rate r for Round Springs? Use the data in the 3rd column of the table and the formula for r from part a.

Maximum flow rate, Round Springs: $r =$ _____ millions of gallons of water per day

g. Calculate the estimated amount of water to flow from each of the springs in an *average* year. Give your answers in millions of gallons per year. (1 year = 365 days)

Amount of water through Big Springs: _____ millions of gallons per year

Amount of water through Round Springs: _____ millions of gallons per year

h. About how many days would it take for one billion gallons of water to flow through each of the springs using the average amount of flow per day? Round your answers to the nearest whole day. *Hint:* Convert one billion into millions of gallons. Then substitute the result into the equation for A, and solve for t.

Big Springs will have one billion gallons of water flow through it in _____ days.

Round Springs will have one billion gallons of water flow through it in _____ days.

1. Extension Exercise—Monitoring Your Stocks

Suppose you decide to go on-line to check the stock that you own in different companies. First you find the following data on Dover Computer Corporation.

Ticker Symbol	Today's High	Today's Low	Last Price	Prior Close	Change	Trade Time
Dover	$42\frac{15}{16}$	$39\frac{1}{2}$	$41\frac{3}{8}$	$43\frac{9}{16}$	↓$2\frac{3}{16}$	12:29 p.m.

a. As of 12:29 p.m. this trading day, what is the difference between today's high price per share and today's low price per share? Show all your work.

b. The Last Price represents the price of a share as of Trade Time 12:29 p.m., and the Prior Close was the price per share at the close of trading on the previous day. Show how subtracting the Last Price from the Prior Close price yields the difference of $2\frac{3}{16}$. Then explain what the arrow indicates in the "Change" column.

_____

_____

_____

c. If you own 1000 shares of Dover, then how much value did your Dover investment lose during the time from the Prior Close price until the Trade Time of 12:29 p.m. today? Do you think this is a cause for concern? Explain.

d. Now observe data from Medicon, another company that you own stock in. What must the "Last Price" be in order to obtain the "Change" of ↑$2\frac{19}{32}$?

Ticker Symbol	Last Price	Prior Close	Change	Trade Time
Medicon	?	$5\frac{1}{4}$	↑$2\frac{19}{32}$	12:29 p.m.

e. If you own 9600 shares of Medicon, then how much value did your Medicon investment gain during the time from the Prior Close price until the Trade Time of 12:29 p.m. today? Do you think this is a cause for celebration? Explain.

2. **Exploration Activity—Bills and Coins**

In the United States, the value of a coin or bill is based on a fraction or multiple of the basic unit, the dollar. A half-dollar and a quarter are so named because they are ½ of a dollar and ¼ of a dollar, respectively.

a. In the table below, write how many coins it takes to equal one dollar and what fraction of a dollar (or 100 cents) each coin is. The first row has been completed for you.

Coin	Number of Coins Needed to Make $1	One Coin is What Fraction of a Dollar?	Simplify the fraction.
Penny	100	$\frac{1}{100}$	$\frac{1}{100}$
Nickel			
Dime			
Quarter			
Half-dollar			

b. Three quarters = _____ pennies

c. Using the information from part *a*, write the fraction of a dollar that represents 75 pennies. Now write the fraction of a dollar that represents three quarters. Show that the two fractions are equal using cross products.

d. One half-dollar has the same value as how many pennies? nickels? dimes? quarters?

1 half-dollar = _____ pennies = _____ nickels = _____ dimes = _____ quarters

Now represent this series of equal values as equivalent fractions. The first fraction is written for you. Prove that all are equivalent fractions by finding the cross products of the first two fractions, the cross products of the second and third fractions, and so on.

$\frac{1}{2} =$

Vanessa works on the floor of a casino selling tokens and chips to customers. The tokens come in rolls, and the rolls are put in *squares* of 25 rolls. A *bucket* is equal to two squares.

e. Complete the table below to determine the value of one roll of each type of token, the value of one square of tokens, and the value of one bucket of tokens. The first row has been completed for you.

Type of Token	Value of Token ($)	No. of Tokens in a Roll	Value of one Roll ($)	Value of one Square (25 rolls) ($)	Value of one Bucket (2 squares) ($)
H	$\frac{1}{2}$	40	$20	$500	$1000
Q	$\frac{1}{4}$	40			
T	$\frac{1}{10}$	100			
N	$\frac{1}{20}$	100			

f. Vanessa invested her hard-earned money and bought 480 shares of stock at $7\frac{3}{8}$ dollars per share. How much did the 480 shares cost? Show your work.

g. In the saying, "Two bits, four bits, six bits, a dollar," follow the pattern to determine how many bits are in one dollar and the value of 2 bits, 4 bits, 6 bits, and 8 bits as a fraction of a dollar. How many quarters equals 2, 4, 6, and 8 bits?

Number of bits in a dollar: _____

No. of Bits	Fraction of a Dollar	×	Number of Quarters in a Dollar	Number of Quarters Equal to 2, 4, 6, or 8 Bits
2		×	4	
4		×	4	
6		×	4	
8		×	4	

3. **Conceptual Exercise—Team Effort**

Doug spends three hours every Saturday morning mowing the lawn while his son, Michael, watches cartoons on TV. The last time that Michael tried to mow the lawn, it took him three hours to complete half the job. Still, Doug figures that if he borrows his neighbor's lawn mower and he and his son work together, this will give his son a sense of accomplishment and a positive work ethic.

a. Since Doug is able to mow the whole lawn in 3 hours, what fraction of the lawn can he mow in 1 hour?

b. If Michael can mow half the lawn in 3 hours, then how long will it take him to mow the whole lawn alone?

c. Using your result from part *b*, what fraction of the lawn can Michael mow in 1 hour?

d. Working together, what fraction of the lawn can Doug and Michael mow in 1 hour? Show all work including the process of finding a least common denominator. Write your answer in simplest form.

e. How long does it take to mow the whole lawn with father and son working together?

f. Why is it helpful to find the fraction of the lawn each person could mow in 1 hour?

_____

_____

_____

g. The concept of finding a common denominator can be visualized with a geometric model. Let the whole lawn be represented by the rectangle below. Split the rectangle into the appropriate number of parts to represent how long it takes Doug to mow the entire lawn. Then shade in the portion of the rectangle to show the part of the job that Doug completes in 1 hour working alone.

h. Now, let the same lawn be represented by another large rectangle given below. Split the rectangle into the appropriate number of parts to represent how long it takes Michael to mow the entire lawn. Shade in the portion of the rectangle to show the part of the job that Michael completes in 1 hour working alone.

i. If Doug and Michael work together, then you can add the fractional amount that each mows in 1 hour separately. Show this geometrically by placing your two models from parts g and h below and find their sum.

Doug   Michael

    +   =

j. What do your answers in part i and part d have in common?

_____

_____

4. **Group Activity—Cubes and Fractions**

Materials Needed:
- Paper to make 2 squares--one with a side length of ½ inch to represent a ½-inch cube, and the other with a side length of ¾ inch to represent a ¾-inch cube
- ruler

A **cube** is a three-dimensional object shaped like a sugar cube where length, width, and height are the same measurement. A cube that is ½ in. x ½ in. x ½ in. is called a half-inch cube (½-inch cube). A cube that is ¾ in. x ¾ in. x ¾ in. is called a three-quarter-inch cube (¾-inch cube). You and your group members will use paper squares to represent one side of a ½-inch cube and one side of a ¾-inch cube.

Work with your group members to answer the following questions.

a. If you line up 13 half-inch-cubes side by side, as illustrated below, then what is the total length of the cubes together? Use a ruler and your ½-inch paper cube (square) to measure the length of 13 half-inches.

Length of 13 half-inch cubes _____ inches

b. How would you use multiplication to find the total length of 13 half-inch cubes? Express the length as an improper fraction and as a mixed number. Show your work.

$$\underline{\hspace{2in}} \text{ inches} = \underline{\hspace{2in}} \text{ inches}$$
$$\text{(improper fraction)} \hspace{1in} \text{(mixed number)}$$

c. If you line up 16 three-quarter-inch cubes side by side, then what is the total length of the cubes together? Use a ruler and your ¾-inch paper cube (square) to measure the length of 16 three-quarter-inches.

d. How would you use multiplication to find the total length of 16 three-quarter-inch cubes? Show your answer as an improper fraction, and then write your answer in simplest form.

Improper fraction: _____ inches   **Simplest form:** _____ inches

How might you build a box with ½-inch cubes? One example is shown below to the right. The bottom layer of the box is shown below to the left.

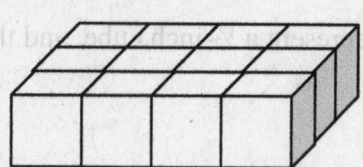

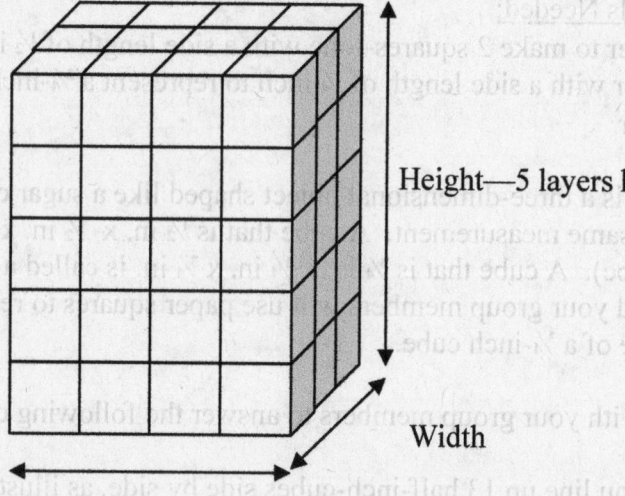

The bottom layer of the box has 3 rows of half-inch cubes and 4 columns or 4 cubes in each row.

One half-inch cube

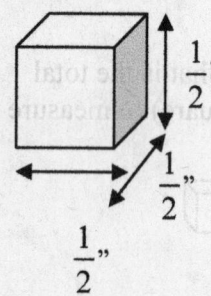

e. Let's now find the length, width, and height of the five-layered box. Use the illustration above to help you find the answer. Express each measurement in terms of number of cubes and number of inches. If a measurement does not result in a whole number of inches, then express your answer as an improper fraction and as a mixed number.

Length: _____ cubes = _____ inches

Width: _____ cubes = _____ inches

Height: _____ cubes = _____ inches

f. The volume of a rectangular box is found by multiplying length × width × height of the box. Calculate the volume of the 5-layered box above using the inch measurements that you found in part *e*. Write the volume in simplest form. If the result is not a whole number, then express your answer as an improper fraction and as a mixed number.

Volume = _____ × _____ × _____ = _____ cubic inches
            length   width   height

1. **Extension Exercise—A Year at MIU**

Suppose you have just started as an exchange student at Maritime International University (MIU). The apartment where you will live is 10 kilometers from campus.

a. If you drive your car from your apartment to the MIU campus, how does travel time depend on average speed (rate)? To find out, complete the table below. Write hours in simplest fractional form. Then give minutes in whole number or decimal form rounded to the tenths place.

$$\text{Time} = \frac{\text{Distance}}{\text{Rate}}$$

Distance (km)	Rate (km/hr)	Time (hours)	Time (minutes)
10	10	1	60
10	20		
10	36		
10	40		
10	45		
10	55		
10	60		
10	70		
10	80		

b. Examine the second and third columns. Every time you double the rate, what happens to the trip time? Answer in complete sentences using examples to support your conclusion.

_____

_____

_____

c. Suppose your average walking pace is 2.7 kilometers per hour. How long will it take you to walk from your apartment to the MIU campus? Show all work and present your answer, the time, in decimal form. Then convert this time to minutes. Round all times to the nearest tenths place.

The fuel economy of your car is 25.6 kilometers per gallon when traveling at a rate of 40 kilometers per hour. Your car's gas tank holds 16.5 gallons.

d. If you are driving at a constant rate of 40 kilometers per hour, what is the total distance (in kilometers) that your car can travel on one tank of gas if all the gas is used?

e. Suppose the price of gasoline is equivalent to $1.99 per gallon. What is the cost of filling an empty tank? Round your answer to the nearest cent.

f. What is the cost of gas per kilometer if your speed is 40 kilometers per hour? Round your answer to the nearest cent.

g. If your first class is at 8 a.m. and you drive at an average speed of 40 kilometers per hour, then at what time should you leave for school each morning? Explain why there might be more than one good answer to this question.

_____

_____

h. If you drive your car from your apartment to the MIU campus each day with no side trips, then how often should you fill your car's gas tank? Round your results to the nearest whole number. *Hint:* You will use your answer from part *d*.

i. If the cost of gasoline stays fixed at $1.99 per gallon for the entire 15-week semester, then about how much will fuel cost over the entire semester? Assume that you do not drive on weekends and that classes occur during the 5-day week. Round results to the nearest cent.

2. **Exploration Activity—Exchange Rates**

In March 2017, the rates of exchange between the United States dollar (USD) and selected foreign currencies were as listed in the table below. Exchange rates can tell you what U.S. dollars are equivalent to in a foreign currency or what a foreign currency is equivalent to in U.S. dollars. For example, on March 10, 2017, 1 USD equaled nearly 7 Chinese yuan (6.91 CNY), almost 20 Mexican pesos (19.68 MXN), or about 115 Japanese yen (115.28 JPY). The values in the table list the exchange rates as units of foreign currency per 1 USD.

The euro/dollar exchange rate can be expressed as $\frac{0.94 \text{ EUR}}{1 \text{ USD}}$, meaning that there are 0.94 euro to every 1 U.S. dollar. This rate can also be written as $\frac{1 \text{ USD}}{0.94 \text{ EUR}}$, meaning that there is 1 U.S. dollar to every 0.94 euro. Assume that all transactions referred to in this activity occur during March 2017 at the rates shown in the table below.

Foreign Currency	Rate Units per 1 USD	Foreign Currency Price of a $500 (USD) TV	Rate USD / unit
Australian dollar - AUD ($)	1.33		
British pound - GBP (£)	0.82		
Canadian dollar - CAD ($)	1.35		
Chinese yuan - CNY (¥)	6.91		
Euro - EUR (€)	0.94		
Japanese yen - JPY (¥)	115.28		
Mexican peso - MXN ($)	19.68		

(Exchange rates recorded on March 10, 2017)

a. Suppose you buy a basic TV for $500 in the U.S. The second column of the table expresses the number of foreign currency units that are equivalent to 1 U.S. dollar (that is, units per 1 USD). Complete the third column of the table by converting the price of the same TV to the currency of each country.

b. If a Canadian citizen buys a car in the U.S. that sells for 20,000 USD, how many Canadian dollars (CAD) must be exchanged to make the purchase in U.S. dollars? Show all work.

c. Suppose a U.S. worker would like to take a vacation to Mexico and decides to set aside one week's pay for spending money while she is on vacation. She earns 11.58 USD per hour (after all payroll deductions) and works a 40-hour week. If she exchanges her entire week's earnings for Mexican pesos, how many pesos will she be able to take with her on vacation? Use the exchange rate given in the table, show all your work, and round your answer to the nearest hundredth.

d. Use the fourth column in the table to list each of the exchange rates as U.S. dollars per unit of foreign currency. For example, 1 USD per 0.94 EUR can be expressed as the ratio $\frac{1 \text{ USD}}{0.94 \text{ EUR}} \approx 1.064$ dollars/euro, meaning there are 1.064 U.S. dollars in 1 euro, or that 1 euro is approximately equal to 1.064 U.S. dollars. Use a calculator as needed. Round results to the nearest thousandth.

e. If you took a vacation on the French Riviera that cost 27,672.30 EUR, how many U.S. dollars would need to be exchanged for the trip? Round your answer to two decimal places.

f. Suppose you took a job in Japan that has a salary of 7,225,800 yen. What would be your salary in U.S. dollars? Round to the nearest dollar.

The table on the first page of this activity is set up for the conversion between U.S. dollars and various foreign currencies. However, you can also calculate the exchange rate of each foreign currency in terms of any other foreign currency. For example, we know that 1 U.S. dollar equals 0.94 EUR, and 1 U.S. dollar equals 0.82 GBP. This implies that 0.94 EUR = 0.82 GBP.

This is an example of cross-exchange rates. The EUR/GBP exchange rate can be calculated by dividing both sides of the equality by 0.82,

$$0.94 \text{ EUR} = 0.82 \text{ GBP}$$
$$\frac{0.94 \text{ EUR}}{0.82} = \frac{0.82 \text{ GBP}}{0.82}$$
$$1.146 \text{ EUR} \approx 1 \text{ GBP}$$

This tells us that 1 British pound (GBP) is approximately equal to 1.146 euros (EUR).

g. Use the ideas developed above to complete the following table of cross-exchange rates. Round to the nearest thousandth.

Currency	units per 1 USD	units per 1 AUD	units per 1 GBP	units per 1 CAD	units per 1 CNY	units per 1 EUR	units per 1 JPY	units per 1 MXN
USD	1							
AUD	1.33	1						
GBP	0.82		1					
CAD	1.35			1				
CNY	6.91				1			
EUR	0.94					1		
JPY	115.28						1	
MXN	19.68							1

3. **Conceptual Exercise—Traveling for Big Brother**

You have been asked to be a consultant for the Big Brother Oil Corporation's mining and coal division. Your task is to examine the most economical way to transport coal to an electric company.

a. Suppose preliminary research shows three possible methods of transportation.

- Trucks traveling 150 miles to the destination
- Freight trains covering 800 miles of railroad to reach the destination
- A barge going for 200 miles on water, then a freight train covering 700 miles of railroad

Do you have enough information to make a recommendation to Big Brother? If yes, go to the next page and write a brief report on the best course of action. Otherwise, state what additional things must be known to make an informed decision.

_____

_____

b. More research reveals the following data on the cost of transporting coal.

- The truck costs $0.17 per ton per mile.
- The freight train costs $0.03 per ton per mile.
- The barge costs $0.02 per ton per mile.

Based on the information gathered so far, which method would you recommend to minimize the cost of moving the coal? Show all work below, then in part *d* write a report to Big Brother Oil Corporation justifying your conclusions.

c. Use the data shown to answer the question below the data.

- The train travels at 70 miles per hour.
- The barge travels at 25 miles per hour.
- The truck travels at 55 miles per hour.
- The train conductor makes 3 stops for a ½ hour break at each stop.
- The trucker makes 3 stops of 15 minutes each.
- The barge conductor travels straight through.
- Time = $\dfrac{\text{distance}}{\text{speed}}$

Assuming that all of the data is true, which of the routes listed in part *b* will take the least amount of time? Show your work, and report the times in hours and minutes.

d. Write a brief report to Big Brother Oil Corporation explaining the most economical way to transport coal to the electric company. Is this the best choice for time reasons? Use your answers from parts *b* and *c* to explain your choice. Give both the pros and cons of your decision.

4.   Group Activity —The Donut Dilemma

The alarm jolts you awake on a Monday morning. It is your turn to bring donut holes to your 8:00 a.m. math class. The local donut shop sells donut holes in boxes of 20, 45, and 60. You have $10 of spending money that must last until Friday, and you are wondering what to purchase at the donut shop. Work with your group members to discuss and answer each of the questions.

a. List what questions you need to consider before you buy any donut holes.

_____

_____

_____

_____

b. Suppose the following data is known:

- One box of 20 donut holes costs $1.99 including tax.
- One box of 45 donut holes costs $3.29 including tax.
- One box of 60 donut holes costs $3.69 including tax.
- There are 33 students registered for your math class.
- On average, 5 students are absent from any class meeting.
- On average, each student eats 3 donut holes.

How many boxes of each size should you buy to make sure that everyone has enough to eat, while at the same time minimizing the cost? Show all work and justify your answer in words using complete sentences.

_____

_____

_____

_____

Chapter 5: Decimals

c. Explain why part b has more than one right answer.

_____

_____

_____

d. What is the maximum number of donut holes that can be bought for $7.00? Show all work and justify your answer in words using complete sentences.

_____

_____

_____

_____

_____

_____

_____

e. How many correct answers are there to part d? In what way is this question different from part b? Explain.

_____

_____

_____

_____

_____

40

1. **Extension Exercise—Cost of Living**

In the working world, you may find it necessary to relocate to get the job you want. When considering job offers in different cities, you should keep in mind that the cost of living in one city is not necessarily the same as the cost of living in another. For example, an annual salary of $42,000 might be plenty to live on in one city but might barely cover housing expenses in another city.

Cost-of-Living Index

City	Index Value
Atlanta, GA	98.7
Boston, MA	148.1
Chicago, IL	118.5
Cleveland, OH	98.7
Denver, CO	110.4
Houston, TX	98.8
Los Angeles, CA	142.3
Miami, FL	111.0
Minneapolis, MN	105.6
New Orleans, LA	97.3
New York, NY	228.2
Oklahoma City, OK	84.6
Pittsburgh, PA	94.0
Seattle, WA	145.1

(*Source:* The Council for Community and Economic Research, 2016)

A cost-of-living index helps us gauge how much more or less expensive it is to live in one city when compared with another. With it, we can find the salary level needed in one city to be equivalent to a given salary in another city using the following proportion:

$$\frac{\text{equivalent salary in City A}}{\text{index value for City A}} = \frac{\text{salary in City B}}{\text{index value for City B}}$$

For example, we can use the given cost-of-living index to find the salary that a worker would need to earn in Chicago to be equivalent to his or her current annual salary of $42,000 in New Orleans. The index value is 118.5 for Chicago and 97.3 for New Orleans. We will let x represent the equivalent salary in Chicago.

$$\frac{\text{equivalent salary in Chicago}}{\text{index value for Chicago}} = \frac{\text{salary in New Orleans}}{\text{index value for New Orleans}}$$

$$\frac{x}{118.5} = \frac{42{,}000}{97.3}$$

$$x \cdot 97.3 = 118.5 \cdot 42{,}000$$

$$97.3x = 4{,}977{,}000$$

$$\frac{97.3x}{97.3} = \frac{4{,}977{,}000}{97.3}$$

$$x \approx 51{,}151$$

Thus, the worker would need to earn $51,151 in Chicago to be equivalent to his or her current salary of $42,000 in New Orleans. Another way to look at this situation is that what this worker can afford with $42,000 in New Orleans would require $51,151 in Chicago.

a. In which city on the given cost-of-living index list is it most expensive to live? In which of these cities is it least expensive to live? Explain your reasoning.

b. Suppose you currently live in Minneapolis, Minnesota, and earn an annual salary of $48,500. You are considering moving to Los Angeles, California. How much would you have to earn in Los Angeles to maintain the same standard of living you have in Minneapolis? Round your answer to the nearest dollar.

c. Suppose you have just graduated from college and have been offered two comparable jobs in different cities. Job A is in Miami, Florida, and pays $46,000 per year. Job B is in Boston, Massachusetts, and pays $56,000 per year. Which job offer would you choose? Explain your reasoning.

d. Suppose that you currently live in Richmond, Virginia. You have been offered two comparable jobs in different cities. Job A is in Cleveland, Ohio, and pays $53,000. Job B is in Atlanta, Georgia, and pays $49,000. Look at the Cost-of-Living Index table and explain why no calculations are necessary to judge which job offer is worth more to you from a cost-of-living standpoint.

2. **Exploration Activity—A Trip to the Ozarks**

Marc is planning a one-week vacation to the Ozarks to canoe, hike, and study geological sites of the area. He is considering renting a Honda Civic, a Ford Taurus, or a Jeep Cherokee but wants to compare rental and gasoline costs before making his decision.

The chart below shows the expected gas mileage (in miles per gallon or mpg), rental costs, and insurance costs for each of the three vehicles.

	Honda Civic	Ford Taurus	Jeep Cherokee
Rental Costs ($ per day)	$38	$42	$63
Insurance Costs ($ per day)	$9	$9	$9
Expected Gas Mileage (mpg) on curvy roads/small towns	32	20	21
Expected Gas Mileage (mpg) on interstate highways	42	29	30

Marc estimates that he will drive a total of about 700 miles during his trip. 230 of those miles will be on the interstate, and the rest will be on curvy, windy roads.

a. Using the data above, find the total number of gallons of gasoline needed for highway driving as well as for curvy-road driving in each of the vehicles.

Hint: Write each mileage as a fraction, $\dfrac{\text{number of miles}}{1 \text{ gallon}}$, and set that fraction equal to the fraction, $\dfrac{\text{\# miles traveled}}{\text{\# of gallons needed}}$, to make a proportion. Fill in the known values and use cross multiplication to find the unknown number of gallons needed. Round results to the nearest tenth.

Car	Curvy Road Proportion	Number of gallons for curvy road driving	Highway Proportion	Number of gallons for highway driving
Honda Civic				
Ford Taurus				
Jeep Cherokee				

Chapter 6: Ratio, Proportion, and Triangle Applications   Key Concept Activity Lab Workbook, *Prealgebra* 8e

b. At a gasoline price of $2.31 per gallon, calculate the cost of gasoline for each vehicle. Then calculate the rental and insurance costs for each vehicle for seven days. Record your answers in the chart below. Round all results to the nearest cent.

Hint: Gas cost = cost of gas per gallon × number of gallons

Honda Civic		Ford Taurus		Jeep Cherokee	
Gas Cost (Curvy road)	Gas Cost (Highway)	Gas Cost (Curvy road)	Gas Cost (Highway)	Gas Cost (Curvy road)	Gas Cost (Highway)
Total Cost of Gas:		Total Cost of Gas:		Total Cost of Gas:	
Rental and Insurance Costs per week:		Rental and Insurance Costs per week:		Rental and Insurance Costs per week:	

c. Which car is the least expensive to rent and drive? Which car costs the most? Other than cost, what do you think Marc should consider when deciding which car to rent?

_____

_____

_____

_____

3. **Conceptual Exercise—The UFO Files**

The Gallup organization conducted a poll asking people about their beliefs in UFO's (unidentified flying objects), extraterrestrial life, and other paranormal subjects.

The results out of 100 people are shown in the table below. In this activity, you will review this data and predict from the sample how many people in the United States are likely to hold similar opinions as the people polled.

Question	Yes	No	No Opinion
1. Have you heard or read about UFO's?	87	13	0
2. Have you ever seen anything you thought was a UFO?	12	87	1
3. Do you think that UFO's have ever visited the earth in some form?	45	39	16
4. Do you think there is life of some form on other planets in the universe?	72	19	9
5. In your opinion, does the U. S. government know more about UFO's than they are telling us?	71	19	10

a. From the information in the chart, set up the following ratios. The first one is done for you.

Question Number	Ratio Stated in Words	Ratio in Numbers
1	number of people who have heard or read about UFO's *to* total number of people polled	87 : 100
2	number of people who think they have seen a UFO *to* total number of people polled	
3	number of people who think UFO's have visited Earth *to* total number of people polled	
4	number of people who think there is life somewhere other than Earth *to* total number of people polled	
5	number of people who think the U.S. government withholds information about UFO's *to* total number of people polled	

b. Suppose that you live in a city of 272,000 people. Based on the ratios from part *a*, set up and solve proportions to determine about how many people in your city would answer "yes" to the five questions on the previous page.

Question Number	Set Up the Proportion	Show Work	Number Who Would Answer "Yes"
1			
2			
3			
4			
5			

c. If it were possible to ask everyone in the United States these questions, do you think the number of people who would answer "yes" would be about the same as the ratios from part *a* would indicate? Explain why or why not.

_____

_____

_____

_____

4. **Group Activity—Eating Well**

Food	Quantity	Protein (grams)	Calories
Egg white	0.6 ounces (1)	4	16
Skim milk	1 cup (8 oz)	8	86
Pink Salmon	3 ounces	21.6	146
Provolone Cheese	1 ounce	7	100
Red Snapper	3 ounces	19.6	89
Ground beef, extra lean	3 ounces	21	170
Yogurt, low-fat plain	1 cup (8 oz)	13	159
Whole egg	1.6 ounces (1)	7	75
Chicken breast, no skin	3 ounces	26.7	142

a. Using the chart above, calculate the ratio of protein grams to ounces of food and the ratio of protein grams to calories. Record your answers in columns 2 and 4 of the chart below. Next, express your answers in terms of the number of protein grams per <u>one</u> ounce and the number of protein grams per <u>one</u> calorie. Round results to the hundredths place.

Food	Ratio of Protein Grams to Ounces of Food	Number of Protein Grams per Ounce	Ratio of Protein Grams to Calories	Number of Protein Grams per Calorie
Egg white				
Skim milk				
Pink Salmon				
Provolone Cheese				
Red Snapper				
Ground Beef, extra lean				
Yogurt, low-fat plain				
Whole egg				
Chicken breast, no skin				

b. Which food has the greatest number of protein grams per ounce? Which food has the fewest number of protein grams per ounce?

Greatest: _____ Fewest: _____

c. Which food has the greatest number of protein grams per calorie? Which food has the fewest number of protein grams per calorie?

Greatest: _____ Fewest: _____

d. Dr. McGeeney prescribed a low fat diet for Julia that has between 80 and 100 protein grams a day. Determine how many protein grams are in the meals below to determine whether Julia is staying within the recommended diet. Round results to the nearest gram. Assume that any vegetables, fruits, and bread have very few (if any) protein grams.

Breakfast:
    2-egg omelet with 2 oz Provolone cheese
    1 glass skim milk (8 ounces)
    2 pieces of toast with jam

Total protein grams for breakfast:

Lunch:
    Grilled chicken salad with 5 ounces skinned chicken breast
    2 bread sticks

Total protein grams for lunch:

Dinner:
    8 ounce ground beef, extra lean
    Baked potato
    2 servings of vegetables

Total protein grams for dinner:

e. Do the meals have the range of total protein grams that the doctor prescribed? If not, what recommendations would you make to Julia?

_____

_____

_____

_____

1. Extension Exercise—Batting Champ

Going into the last day of the major league season, Babe Boggs and Kirby Lockett had identical batting records of 200 hits in 600 at bats. On the last day of the season, Boggs had 7 hits in 8 times at bat, while Lockett had 9 hits in 12 times at bat. Who do you think won the batting title?
Note: The batting title goes to the hitter with the highest ratio of "hits" to "at bats."

a. Calculate the identical batting average of Boggs and Lockett just before the last day of the season. The batting average is the ratio (fraction) of "hits" to "at bats." Express this ratio as a fraction in lowest terms, a decimal rounded to the thousandths place, and a percent.

$$\frac{hits}{at\ bats} =$$

b. What is the practical meaning of the batting average when expressed as a percent?

_____

_____

_____

c. Calculate Boggs' and Lockett's batting average just for the last day of the season. Express the ratio as a fraction in lowest terms, a decimal rounded to the thousandths place, and a percent. Who had the most hits on the last day? Who had the higher batting average if we just use the last day statistics?

<u>Last Day Statistics</u>

Boggs: $\dfrac{hits}{at\ bats} =$ Lockett: $\dfrac{hits}{at\ bats} =$

Most Hits:

Highest Average:

d. Which player do you think won the batting title? Explain using complete sentences.

_____

_____

_____

_____

e. Calculate the final batting average for each player at the end of the baseball season. Express the ratio as a fraction in lowest terms, a decimal rounded to the thousandths place, and a percent. Who had the most hits at the end of the season? Who had the higher batting average at the end of the season?

<u>Final Season Statistics</u>

Boggs: $\dfrac{hits}{at\ bats} =$ Lockett: $\dfrac{hits}{at\ bats} =$

Most Hits:

Highest Average:

f. Who won the batting title? Explain using complete sentences.

_____

_____

g. Let's return to the last day of the season when Boggs had 7 hits in 8 at bats, while Lockett had 9 hits in 12 at bats. The table below shows how Lockett progressed through the last game. Fill in the last three rows, giving his season batting average after each "at bat" as a ratio, a decimal, and a percent. Round the decimal to the thousandths place.

Time at Bat	1st	2nd	3rd	4th	5th	6th	7th	8th	9th	10th	11th	12th
Outcome	Hit	Hit	Hit	Out	Hit	Hit	Hit	Hit	Out	Hit	Out	Hit
Current Average (ratio)												
Current Average (decimal)												
Current Average (percent)												

2. **Exploration Activity — A Diverse India**

In 2017, India celebrated its 70th anniversary of independence. The country was created from 600 states into a diverse nation with many ethnic groups that speak 18 official languages and practice 7 major religions. India is a study in contrasts when it comes to food, education, and wealth. Although the country is agriculturally self-sufficient, nearly 30% of children under 5 are malnourished. The literacy rate jumped from 12% in 1947 at the time of its independence from Britain to 74% in 2011, but there is still a wide gap between men and women and between those who live in urban and rural areas.

In March of 2017, the United Nations estimated India's population as 1,337,756,334, and the national currency, the rupee, had the following exchange rate with the U.S. dollar:
1 dollar = 65.78 rupees. We will use this data in this activity.

a. The World Bank estimates that 32.7% of Indian people live on less than $1.25 per day. Calculate the number of people living in India who try to survive each day with less than $1.25 to spend. Round your answer to the nearest whole.

b. Suppose a man manages to save $1.25 of extra spending money per month. If a basic pair of shoes cost 700 rupees, then how many months will it take him to save enough money to buy the shoes? Round your answer to one decimal place.

c. A woman who makes 2,500 rupees per month has the goal of saving enough money in one year to buy a television worth 21,000 rupees. How much money must she save each month in order to reach her goal? How much money does she have left each month to pay bills and live on? What is this amount in U.S. dollars? Round this amount to the nearest cent.

   Monthly Savings (rupees):

   Monthly (rupees) to live on: Monthly ($) to live on:

d. India's two largest religions are Hinduism and Islam. If Hindus outnumber Muslims 5.7 to 1 and 80% of the country is Hindu, then what percentage is Muslim? Round the percent to the nearest whole percent and populations to the nearest whole.

   Percent of Muslims in Population:

   Number of Hindus in Population:

   Number of Muslims in Population:

e. For each year listed in the table, compute India's share of world population as a percent. Round each percent to two decimal places. Describe any trends you notice.

Year	Population of India	World Population	India's Share of World Population (%)
2017	1,342,512,706	7,515,284,153	
2016	1,326,801,576	7,432,663,275	
2010	1,230,984,504	6,929,725,043	
2000	1,053,481,072	6,126,622,121	
1990	870,601,776	5,309,667,699	
1980	697,229,745	4,439,632,465	
1970	553,943,226	3,682,487,691	
1960	449,661,874	3,018,343,828	

(*Source:* Worldometers)

f. For each pair of years listed in the table, use the data for Population of India given in the previous table to compute the amount of increase in India's population from the first year listed to the second year listed. Then calculate the associated percent of increase. Round each percent to two decimal places.

From Year	To Year	Amount of Increase in India's Population	Percent of Increase
2016	2017		
2010	2017		
2000	2017		
1990	2017		
1980	2017		
1970	2017		
1960	2017		

3. **Conceptual Exercise —Where's the Beef?**

A supervisor at a meat packing plant needs to maintain a certain level of quality control for the product due to federal law. Suppose that the supervisor is in charge of marinated beef. In that department, workers place up to 500 pounds of meat into a tumbling machine and add a certain amount of marinade. To maintain quality, the final product cannot have more than a 20% increase in weight due to adding marinade.

a. An employee named John uses his memory of past experiences to estimate the amount of marinade added. John has just processed 200 pounds of beef, and the final product of marinated meat weighs 234 pounds. If a meat inspector makes a surprise visit, will the inspector give the company a violation? Show all work, and write your answer in complete sentences.

_____

_____

_____

b. Suppose that 250 pounds of newly arrived beef (called the green product) is placed in a tumbling machine, and some marinade is added before packaging the beef (the final product) for shipment. What is the maximum possible weight of the marinated meat, if it is to stay within federal guidelines? Show all work and write your answer in complete sentences.

_____

_____

_____

_____

c. The meat inspector arrives at the plant and asks the supervisor to explain what method the plant employees use to calculate the 20% rule. This is the process that he describes:

1) Weigh the green product and record the value in pounds.
2) Take the weight in pounds and move the decimal point of this number one place to the left.
3) Double the number from Step 2 to arrive at the maximum increase in weight.
4) The final product is discarded if it weighs more than the green product plus the maximum increase just found in Step 3.

Does this procedure correctly meet the 20% rule required by federal law? Support your written answer by showing how the calculations work on 250 pounds of green product meat.

_____

_____

_____

d. The meat inspector wants the supervisor to create a percent change formula that involves using the words *final product* and *green product*. An employee should be able to substitute the weight of the green product and final product into the formula, calculate the percent change, and then compare the value with 20%. Write this percent change formula below.

e. Test your percent change formula from part *d*. Use the formula in the following two situations.

- A plant employee weighs a fresh piece of meat at 230 pounds before mixing it with marinade. After the marinade is added, the weight of the beef is 272 pounds. Would the final product pass inspection under federal law?

- The same employee weighs a second portion of meat at 235 pounds before adding the marinade. After mixing it with marinade, the new weight is 284 pounds. Would the final product pass inspection under federal law?

4. **Group Activity—M&M's®**

In the 1930s, many American stores did not stock chocolates during the summer because, without widespread air conditioning, the chocolate tended to melt and sales declined. In 1940, Forrest E. Mars, Sr., set out to develop a chocolate candy that could be sold year-round without the melting problem. So he formed a company in Newark, New Jersey, to make bite-sized melt-proof chocolate candies encased in a thin sugar shell. Thus M&M's® Plain Chocolate Candies were born! The first M&M's® were made for the U.S. military in 1941.

The first Plain M&M's® colors were brown, green, orange, red, violet, and yellow. Violet was replaced by tan in 1950. When the safety of a particular type of red food coloring was publicly questioned in 1976, red was completely eliminated from the M&M's® color mix to avoid alarming consumers, even though the coloring in question was never used in M&M's®. After an 11-year hiatus, red returned to the color mix in 1987. In 1995, over 10 million Americans responded to a marketing campaign asking for help in choosing a new M&M's® color. Given the choices of blue, pink, purple or no change, 54% of the respondents chose blue, and it replaced tan. At that time, the new color mix of Plain M&M's® became blue (10%), brown (30%), green (10%), orange (10%), red (20%), and yellow (20%).

According to Mars, Incorporated, the current distribution of M&M's® colors depends on which of two different U.S. factories produces them. The M&M's® factory in Hackettstown, New Jersey, produces M&M's® with the following official color distribution:

Blue	Brown	Green	Orange	Red	Yellow
25.0%	12.5%	12.5%	25.0%	12.5%	12.5%

a. Open a single-serving pouch of M&M's® Plain Chocolate Candies. Count the blue, brown, green, orange, red, and yellow M&M's® in your pouch and record the data in table below.

Blue	Brown	Green	Orange	Red	Yellow

b. What was the total number of M&M's® in your pouch?

c. Find the percentage of each color in your pouch and record them in the table below. Round each percent to one decimal place.

% Blue	% Brown	% Green	% Orange	% Red	% Yellow

d. How do the percents you calculated in part *c* compare to the current official color distribution at the Hackettstown plant?

Chapter 7: Percent

e. Combine your group's color counts with those of the other groups in your class. Use the following table to organize your data. Find candy totals for each color and the overall total number of M&M's® tallied by all of the groups.

Group	Blue	Brown	Green	Orange	Red	Yellow	Total
1							
2							
3							
4							
5							
6							
7							
8							
9							
10							
TOTALS							

f. Find the percentage of each color for the combined data in part *e*, and report your results in the table below.

% Blue	% Brown	% Green	% Orange	% Red	% Yellow

g. How do these percentages for the combined data compare to the official color mix at the Hackettstown plant? Do the combined data give color percentages that are more similar or less similar to the official percentages than those you computed for a single pouch in part *c*? Explain.

1.   Extension Exercise—Winning BigBucks Lottery

Steve and Kate are wondering what 6 numbers to choose for Saturday's lottery game called BigBucks. This is a 6/49 lottery, meaning you play by choosing 6 different numbers between 1 and 49 inclusive and hope that your selection matches the winning combination (randomly picked by the State Lottery Commission).

Steve uses a number obtained by calling 1-900-PSYCHIC to try and win the $2,674,763 jackpot. Kate decides on the following 6-number combination, 1, 2, 3, 4, 5, 6.

Her logic is that since any 6-number combination is equally likely to occur, the above pick has as good a chance of winning as any other combination.

a. Let's start by exploring the six-number combination you must select to play a 6/49 lottery. How many possible numbers are available when choosing the first number?

b. Once the first number is chosen, how many numbers do you have available when choosing a second number (different from the first number)?

c. How many ways are there to choose the first two numbers? Write your answer as a product of two numbers, and then compute the result.

d. After the first two numbers are chosen, how many numbers are available when choosing a third number (different from the first and second numbers)?

e. How many ways are there to choose the first three numbers? Write your answer as a product of three numbers, and then compute the result.

f. Knowing that there are $49 \cdot 48 \cdot 47 = 110{,}544$ ways to choose the first three numbers, how many ways can you choose the entire winning six-number combination? Write your answer as the product of six numbers, and then compute the result.

In the last six questions, we have treated the order of appearance of each number as being important (different). For example, we have counted combinations such as
48 7 16 as being different from 7 16 48.

In fact, when playing the lottery, the order in which the numbers appear is not important, so we have over-estimated the number of ways to choose the entire six-number combination necessary to win BigBucks lottery. Let's try to figure out what value must be divided out, so that we are counting only the different combinations without regard to order.

g. Answer the following questions to find how many ordered arrangements (or permutations) of 6 numbers exist.

- Given six distinct numbers, how many ways can you choose the first number? _____
- Having chosen the first number, how many ways can you choose the second number?

- Having chosen the first and second numbers, how many ways can you choose the third number?

- Having chosen the first, second, and third numbers, how many ways can you choose the fourth number?

- Having chosen the first, second, third, and fourth numbers, how many ways can you choose the fifth number?

- Having chosen the first, second, third, fourth and fifth numbers, how many ways can you choose the sixth number?

- Altogether, how many ways can you order six numbers?

h. Find the quotient of the product from part f divided by the product from part g. The result is all the possible six number combinations that have an equally likely chance to occur.

Number of possible outcomes = $\dfrac{\text{product from part } f}{\text{product from part } g}$ =

i. If a favorable outcome is selecting the winning number with your ticket, then how many favorable outcomes exist?

Number of favorable outcomes from your ticket:

j. Use the probability ratio to find the chance (probability) of your ticket winning the BigBucks lottery game.

Probability of winning = $\dfrac{\text{Number of favorable outcomes}}{\text{Number of possible outcomes}}$ =

2. **Exploration Activity—What Is Average?**

Mr. Smith is ready to give back graded midterm exams to his class of 32 students. A student asks him, "How did the class do?" Mr. Smith displays the following chart on the overhead projector.

56	72	74	67	74	92	81	76
82	90	75	98	61	88	65	94
63	58	82	50	77	83	78	86
69	82	61	75	85	78	61	72

a. Mr. Smith asks you to complete the table given below.

Class Intervals (Exam Scores)	Class Frequency (Number of Students within given grade range)
50–59	
60–69	
70–79	
80–89	
90–99	
TOTAL Number of Students	

b. Next Mr. Smith asks you to construct a histogram from the table in part *b*. Choose an appropriate scale for the vertical and horizontal axes.

Number of Students

Exam Scores

c. Suppose that Mr. Smith uses the following system to convert numerical scores to letter grades:

   A: 90-100   B: 80-89   C: 70-79   D: 60-69   F: 0-59

   Answer the questions below and round each to the nearest whole.
   - What percent of the students in the class received an *A*?

   - What percent of the students in the class received a *B*?

   - What percent of the students in the class received a *C*?

   - What percent of the class passed (60-100) the midterm?

   - What percent of the students failed (0-59) the midterm?

d. Use the information from parts *a* and *c* to create a pie graph (circle graph) of the percentages of A's, B's, C's, D's, and F's in the class.

e. For the given set of exam scores,
   - find the mean score of the class. Round to the nearest whole.
   - find the median score of the class.
   - find the mode score of the class (there may be more than one).

f. Which measures of central tendency best describe the exam scores of the class? Explain using complete sentences.

_____

_____

_____

3. **Conceptual Exercise—Lightning Strikes**

Suppose we want to compare the probability of winning a 6/49 lottery, with the probability of a randomly selected American being struck by lightning, over a one-year time period. According to Worldometers, the resident population of the United States is approximately 325,000,000. According to the National Oceanic Atmospheric Administration, an average of 310 people are struck by lightning in the United States each year.

a. What is the probability that a randomly selected American resident will be struck by lightning this year? State the probability as a fraction, a decimal rounded to a single non-zero digit, and a percent. Use complete sentences to explain your answers.

Fraction:

Decimal:

Percent:

_____

_____

_____

b. Do you think that every resident of the United States has the exact same probability of being hit by lightning? Use complete sentences to explain your answer.

_____

_____

_____

_____

c. If you round the chances of winning the lottery to 1 in 14 million, then how many more times likely are you to be hit by lightning than to win the lottery? Show all work and explain your answer using complete sentences.

4. Group Activity—Laundry Blues

Tom woke up on a rainy Monday morning with nothing to wear for class. He forgot to do his laundry yesterday. It turns out that his roommate John actually did some of his laundry for him. There are 4 clean shirts and 3 clean pairs of pants hanging in the closet. Tom thinks that he doesn't have enough choices to mix and match into a suitable outfit to wear, but John says that there are 12 possible outfits to choose from, and at least one of these outfits must match!

a. Let's define the 4 shirts as S#1, S#2, S#3, and S#4, and the 3 pairs of pants as P#1, P#2, and P#3. Use a tree diagram to list the 12 possible outfits that Tom can choose from.

b. Suppose there are 8 possible combinations that Tom would consider suitable. If he walks into the dark closet and randomly chooses one pair of pants and one shirt without being able to distinguish colors or styles, then what is the probability that Tom has picked an outfit that he considers to be suitable? Show all work, and state your answer as a fraction, a decimal rounded to the hundredths place, and a percent.

Fraction: Decimal: Percent:

c. What is the probability that in the dark closet Tom has randomly picked out an outfit that is not suitable? Show all work, and state your answer as a fraction, a decimal rounded to the hundredths place, and a percent.

Fraction: Decimal: Percent:

d. Assume that S#1 is Tom's favorite shirt. What is the probability of randomly picking this shirt in the dark closet? Show all work, and state your answer as a fraction, a decimal rounded to the hundredths place, and a percent.

Fraction: Decimal: Percent:

e. Suppose there is one shirt with two buttons missing and 1 pair of pants with a broken zipper, what is the probability that Tom has picked:

- both these items?

   Fraction: Decimal (to nearest hundredths place):   Percent:

- one of them?

   Fraction: Decimal (to nearest hundredths place):   Percent:

f. Tom finally chooses an outfit. While getting dressed, he listens to the radio and hears that there is a 75% chance of rain. Does this mean that rain will fall from the sky 75% of the time that day? Support your answer using complete sentences.

_____

_____

_____

_____

_____

_____

1. **Extension Exercise— Survival of the Hatchlings**

There are about 29,000 - 40,000 adult female Leatherback sea turtles in the world. On average, a female turtle lays eggs every 2 - 3 years. She nests 4 or 5 times in one season, laying about 90 eggs per nest. Approximately 40 percent of the eggs hatch, and 1 in 2500 hatchlings survive to adulthood.

a. Since female Leatherbacks lay eggs on average every 2 to 3 years, we could roughly guess that in one year about one-half to one-third of the females are nesting. Estimate the fewest number of females nesting in one year, assuming that $\frac{1}{3}$ of the lowest estimate, 29,000 adult females, nest in that year. Round your answer to the nearest whole number. Show your work and answers below.

Least number of females nesting in a year: _____

b. Estimate the greatest number of females nesting in a year, assuming that $\frac{1}{2}$ of the high estimate, 40,000 adult females, nest each year. Show your work and answers below.

Greatest number of females nesting in a year: _____

c. Use the information from parts *a* and *b* to show the probable range of Leatherback female turtles nesting each year.

From _____ to _____ Leatherback female turtles nest each year.

d. Determine a range of how many eggs are expected to hatch from one female Leatherback turtle in one season (if this is her nesting year). Show your work.

Hint: # of nests per year for one turtle × # of eggs per nest × 40% = # of turtles hatched per one female Leatherback turtle

From _____ to _____ hatchlings by one nesting female turtle each year.

e. Review the information in steps c and d, and record the range of expected total number of hatchlings in any year.

   Hint: total # of hatchlings = # of female turtles nesting × # of hatchlings per turtle.

   From _____ to _____ Leatherback turtle eggs are hatched each year.

f. If the survival rate of hatchlings is 1 : 2500, then what is the range of numbers of each year's hatchlings expected to survive to adulthood? Set up and solve the proportions to answer the question. Round each answer to the nearest whole.

   From _____ to _____ Leatherback hatchlings are expected to survive to adulthood.

g. Mature Leatherbacks measure anywhere from 1.2 to 1.9 meters long and weigh from 200 to 500 kilograms. The heaviest Leatherback ever recorded was 916 kilograms. Convert these metric measurements and weights to the approximate number of units in the U.S. system. Rewrite the first two sentences of this part using U.S. system equivalents. Round your U.S. units to the nearest tenth for length and to the nearest whole number for weight. Show the unit fractions and set-ups in your conversions.

   Hint: 1 meter ≈ 3.28 feet (≈ means "approximately equals"); 1 kilogram ≈ 2.20 pounds

   1.2 meters ≈ _____

   1.9 meters ≈ _____

   200 kilograms ≈ _____

   500 kilograms ≈ _____

   916 kilograms ≈ _____

2. **Exploration Activity— Fence Me In**

Suppose you just bought a house that sits on a rectangular lot with an area of 7200 square feet. The house is in good shape, but you want to enclose the borders with a fence. You know that the lot is 7200 square feet, and yet you still don't know the actual length or width of the lot.

a. Use the given data below to complete the table and list some possible dimensions for your 7200 square foot lot. Draw a picture of the situation to help you visualize the problem.

$Area = Length \cdot Width$
$Perimeter = 2 \cdot (Length + Width)$

Table with some given lengths:

Area	Length	Width	Perimeter
7200	40		
7200	60		
7200	72		
7200	80		
7200	90		
7200	100		
7200	120		
7200	180		

b. In the above table, what happens to the width as the length increases? Answer using complete sentences, and explain any patterns you observe.

_____

_____

_____

c. In the table above, what happens to the perimeter as the length increases? Answer using complete sentences, and explain any patterns you observe.

_____

_____

_____

d. Suppose you want to enclose the entire perimeter (border) of your property with 6 foot high fencing that cost $5.00 per foot of length purchased. The total length of fence you need depends on the perimeter. If you could choose any length/width pair from the previous table, what dimensions would minimize the cost of fencing in your property? Justify your answer using complete sentences.

_____

e. Do you think there are dimensions not listed in the table that would decrease the cost of fencing even more than your answer to part *d*? If yes, approximate the length and width. Explain your answer using complete sentences.

_____

f. Let's investigate by looking at another table that starts with a length of 80 ft and goes up in 1-foot increments to a length of 90 feet. Fill in the columns for width, perimeter, and fencing cost using the equations below. Round results to the nearest hundredth.

$$Width = \frac{7200}{Length} \qquad Perimeter = 2 \cdot (Length + Width) \qquad Fencing\ Cost = \$5 \cdot (6 \cdot Perimeter)$$

Area	Length	Width	Perimeter	Calculations	Fencing Cost
7200	80				
7200	81				
7200	82				
7200	83				
7200	84				
7200	85				
7200	86				
7200	87				
7200	88				
7200	89				
7200	90				

g. Which perimeter gives the lowest fencing cost? If width values had to be rounded to the nearest foot, which dimensions would result in the lowest fencing cost?

_____

3. **Conceptual Exercise—Converting Doses**

The pharmacist has in stock 20 bottles of amoxicillin with 100 capsules in each bottle plus a partially full bottle with 25 capsules. Each capsule contains 250 milligrams of amoxicillin.

Information: 1 kilogram = 1000 grams.   1 gram = 1000 milligrams.

a. How many total milligrams of amoxicillin does the pharmacy have on hand? What is that amount in grams and in kilograms? Round results to the nearest hundredth place.

amoxicillin on hand: _____ mg = _____ g = _____ kg

b. One kilogram equals how many milligrams? *Hint:* Use your answers from part *a* if you need a reference and set up a proportion to answer the question.

1 kg = _____ mg

c. Doctor Alvarez prescribed 750 milligrams of amoxicillin per day for seven days. How many total milligrams did the doctor prescribe? What is that amount in grams?

amoxicillin prescribed: _____ mg = _____ g

d. Since there are 250 mg of amoxicillin in each capsule, how many capsules should the pharmacist put in the bottle to fill the prescription described in part *c*?

_____ capsules

e. If the prescription from part c is followed, how many capsules should be taken in one day?

_____ capsules per day

f. The pharmacist must make up a label to tell the patient how many capsules to take and when. If the capsules are to be taken in equally spaced intervals over 24 hours, then what instructions should be written on the label? *Hint:* Tell the patient how many capsules to take at which points during the day.

_____

_____

_____

_____

_____

_____

g. How much amoxicillin is in stock after the prescription (from part c) is filled? Express your answer in milligrams as well as grams.

amoxicillin in stock: _____ mg = _____ g

4.   Group Activity—Composting

To keep your garden soil rich with organic matter, you decide to keep your yard waste in a compost pile. Building a bin using extra wood and wire will promote faster decomposition. A good compost bin is shaped like a cube and has a volume ranging from 27 to 125 cubic feet ($ft.^3$). Work with your group members to answer the following questions.

Materials: paper, scissors, ruler

a. A die that you might toss in a board game is an example of a cube. How many numbered faces (sides) does it have?

b. Each face on a cube is an identical square. Represent the compost bin by drawing a wire frame cube that shows all the edges (lines) that would make up the 6 faces.

c. What is the range of length measurements for each identical face (side), if we are to stay within the recommended limits for the volume of the compost bin? *Hint:* ft × ft × ft = ft^3.

d. Let's make two scaled-down models of the smallest and largest possible bins (with no lid). Use a piece of paper along with scissors and a metric ruler to construct a scaled down model of our cubic bin, with a measure of 1 inch being equivalent to 1 foot of the real bin. To model the smallest recommended bin, use the dimensions, 3 in × 3 in × 3 in, giving a volume of 27 cubic inches (in^3). Follow the steps below.

1.) Draw a square with each of the 4 sides having a length of 9 in.

2.) From each of the 4 corner points (vertices), measure and mark 3 inches in the horizontal direction and 3 inches in the vertical direction.

3.) Use your ruler to draw a 3 inch by 3 inch square in each of the 4 corners.

4.) Use your scissors to cut out each 3 inch by 3 inch square in each of the 4 corners.

5.) Fold up all of the sides to form an open box that will model our compost bin.

Next, model the largest possible bin using the dimensions, 5 inches by 5 inches by 5 inches giving a volume of 125 cubic inches.

e. Which of the following two choices would be the most economical for composting? Answer using complete sentences and support your answer with mathematical reasoning.

- 1 compost bin with a volume of 125 cubic feet

OR

- 4 compost bins each with a volume of 27 cubic feet

f. Suppose you only have enough material to make a compost bin measuring $3\frac{1}{2}$ ft $\times 3\frac{1}{2}$ ft $\times 3\frac{1}{2}$ ft. How many cubic feet of lawn waste could it hold?

g. Suppose you can carry $\frac{1}{2}$ cubic foot of lawn waste in each full shovel. If you end up with 80 full shovels of lawn waste, will the bin that you constructed in part f overflow? Explain.

h. What are the dimensions of a cubic compost bin that has a volume of 64 cubic feet? Explain how you found these dimensions.

1. **Extension Exercise—Shipping Box of Maximum Volume**

According to the United States Postal Service, most parcels may measure no more than 108 inches in combined length and girth. For rectangular shipping boxes, this means that the sum of the box's length, width, and height, $L + W + H$, must not exceed 108 inches.

An interior designer wants to mail a custom-made curtain rod to a client and needs a rectangular cardboard box of maximum volume for shipping the rod, along with plenty of package filler to cushion it in transit. The curtain rod is 63 inches long. The interior designer decides to allow an extra inch in length so that the curtain rod will easily fit in the cardboard box.

a. If the length, L, of the box is 64 inches, how long can the combined width and height of the box be so that all three dimensions add up to 108 inches?

$W + H = $ _____

b. Solve the equation from part *a* for W.

c. The volume of a rectangular box is given by the formula $V = LWH$. Substitute the value for L and the expression for W (from part *b*) into the formula for volume to express volume V in terms of the variable H only. (In computing the volume, we are ignoring the thickness of the box's cardboard walls. The actual volume will be slightly less because of the space the cardboard itself takes up.)

$V = $ _____

d. Write the expression from part *c* in another form by multiplying the factors.

$V = $ _____

e. Compute V for each of the values of H below. Record your answers in the table.

H in inches	V in cubic inches
15	
18	
21	
24	
27	

Chapter 10: Exponents and Polynomials

f. Examine the table from part *e* to find the height that gives the greatest volume for the box. What is the volume of the box with that height? Include correct units with your answers.

height $H = $ _____ volume $V = $ _____

g. Identify the value of H in the previous table that is closest to but less than the value of H from part *f* and the value closest to but greater than that value of H from part *f*: _____. Including these two values of H, create a new table for all whole-number values of H between them.

H in inches	V in cubic inches

What is the greatest volume in this table? $V = $ _____. What are the dimensions of the box of greatest volume that fall within the required specifications? Include correct units with your answers.

$L = $ _____ $W = $ _____ $H = $ _____

2. **Exploration Activity—Computing Distances**

When a rock is dropped from the edge of a cliff that is 400 feet above the ground, the approximate distance (in feet) of the rock above the ground after t seconds can be computed by the formula,

$$d = (-4t + 20)(4t + 20).$$

a. Multiply the binomial factors in the previous formula to obtain a different form of this equation.

$d =$ _____

b. Graph $d = (-4t + 20)(4t + 20)$ for the first 5 seconds after the rock is dropped.

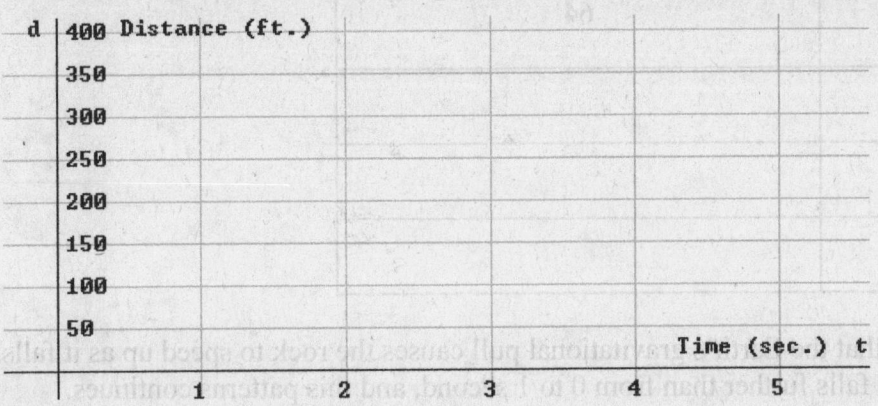

c. Find the formula for the distance (in feet) that the rock falls after t seconds. (*Hint:* The formula you have gives the rock's distance above the *bottom of the cliff.* The distance that the rock has fallen is measured from the *top of the cliff.*)

$d =$ _____

d. How far has the rock fallen in 5 seconds? Show or explain how you came up with your answer.

e. How far is the rock from the ground after falling for 5 seconds? Show or explain how you came up with your answer.

f. Fill in the missing data in the chart below. Calculate and record in the second column how far the rock has fallen at various points in time.

Time (in sec)	Distance Rock Falls (in ft)
0	0
1	16
2	64
3	
4	
5	

Notice from the table that the Earth's gravitational pull causes the rock to speed up as it falls, so from 1 to 2 seconds, it falls further than from 0 to 1 second, and this patterns continues.

3. Conceptual Exercise—Deaths from Motor Vehicle Crashes

Each year the Insurance Institute for Highway Safety collects data on motor vehicle crash deaths in the United States, categorized by passenger vehicle occupants, pedestrians, motorcyclists, bicyclists, and large truck occupants.

According to data from the Insurance Institute for Highway Safety, a polynomial that represents the annual number of deaths of **passenger vehicle occupants** over the period from 2003 through 2013 is
$43x^3 - 621x^2 + 872x + 31{,}964$ where $x = 0, 1, \ldots, 10$ represents the years 2003, 2004,…, 2013.

Also based on data from the Insurance Institute for Highway Safety, a polynomial representing the annual number of deaths due to **all other types** of motor vehicle crashes (including pedestrians, motorcyclists, bicyclists, and large truck occupants) over the period from 2003 through 2013 is
$19x^3 - 292x^2 + 1181x + 10{,}576$ where $x = 0, 1, \ldots, 10$ represents the years 2003, 2004,…, 2013.

a. Use the polynomials to complete the table below. Find the number of passenger vehicle occupant deaths per year and the number of deaths due to all other types of motor vehicle crashes over the period of 2003 through 2013 by evaluating each polynomial at the given values of x. Next, discuss how to use the data in the table to calculate the total number of motor vehicle crash deaths for each of the given years and then do so.

Year	x	Number of Deaths of Passenger Vehicle Occupants	Number of Deaths Due to All Other Types of Motor Vehicle Crashes	Total Number of Motor Vehicle Crash Deaths
2003	0	31,964	10,576	42,540
2005	2	31,568	11,922	43,490
2007	4	28,268	11,844	40,112
2009	6	24,128	11,254	35,382
2011	8	21,212	11,064	32,276
2013	10	21,584	12,186	33,770

Chapter 10: Exponents and Polynomials

b. Use the two given polynomials to find a new polynomial that represents the **total number** of motor vehicle crash deaths per year and record it as P on the blank below. Evaluate this new polynomial P for x = 0, 2, 4, 6, 8, and 10 and record your results in the table. Copy over your answers from the last column in part a.

New polynomial, P = _____

x	P evaluated at given x-value	Data from last column in part a
0		
2		
4		
6		
8		
10		

c. Compare the values in the second and third columns of the table from part b. What do you notice? What can you conclude?

4. Group Activity—Worldwide Internet Users

The number of people around the world who use the Internet continues to grow and grow. However, the worldwide population continues to increase as well, so it may be difficult to understand trends in the growth of Internet use with absolute numbers alone. In this activity, your group will explore the growth of worldwide Internet users as compared to the growth of worldwide population.

a. The number of worldwide Internet users (in millions) x years after the year 2000 is given by the polynomial $4.8x^2 + 104x + 431$ for the years 1995 through 2015. Work with your group to complete the following table. The first line has been completed for you.

Year	x	$4.8x^2 + 104x + 431$ evaluated at x	Number of worldwide Internet users written in **standard form**
1995	−5	$4.8(-5)^2 + 104(-5) + 431 = 31$ million	31,000,000
1998	−2	$4.8(-2)^2 + 104(-2) + 431 = 242.2$ million	242,200,000
2001	1	$4.8(1)^2 + 104(1) + 431 = 539.8$ million	539,800,000
2004	4	$4.8(4)^2 + 104(4) + 431 = 923.8$ million	923,800,000
2007	7	$4.8(7)^2 + 104(7) + 431 = 1394.2$ million	1,394,200,000
2010	10	$4.8(10)^2 + 104(10) + 431 = 1951$ million	1,951,000,000
2013	13	$4.8(13)^2 + 104(13) + 431 = 2594.2$ million	2,594,200,000

b. The following table lists the approximate total worldwide population, according to the United States Census Bureau International Data Base, for each year given. Work with your group to complete the table by writing each population in standard form.

Year	World population	World population written in **standard form**
1995	5.7 billion	5,700,000,000
1998	5.9 billion	5,900,000,000
2001	6.2 billion	6,200,000,000
2004	6.4 billion	6,400,000,000
2007	6.6 billion	6,600,000,000
2010	6.9 billion	6,900,000,000
2013	7.1 billion	7,100,000,000

c. Begin working on the table shown below by copying your results for the "Number of worldwide Internet users written in **standard form**" (last column of the table in part *a*) and for "World population written in **standard form**" (last column of the table in part *b*). Working with your group, use these values to complete the table by finding what percent of the world population were Internet users in each year listed. Be sure to show your work. Round each percent to the nearest tenth of a percent.

Year	Number of worldwide Internet users written in **standard form**	World population written in **standard form**	Percent of world population that were Internet users
1995	31,000,000	5,700,000,000	$\dfrac{31{,}000{,}000}{5{,}700{,}000{,}000} \approx 0.05\%$ 0.5%
1998			
2001			
2004			
2007			
2010			
2013			

d. Describe any trends that you see in the data table in part *c*.

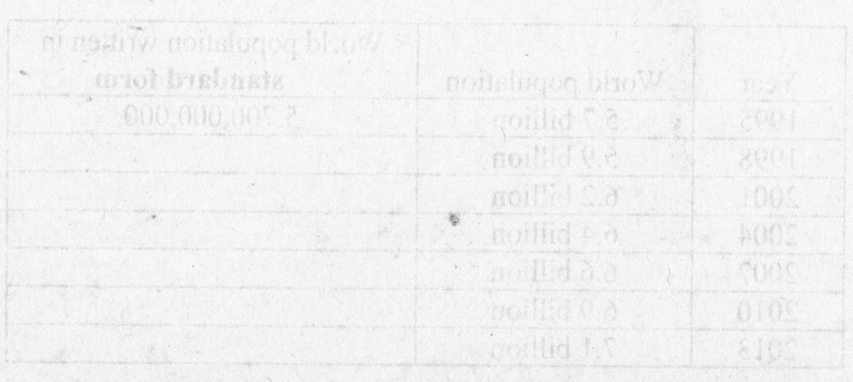